ELECTRICITY

SCIENCE
ENERGY
INVENTIONS

Amanda Bennett

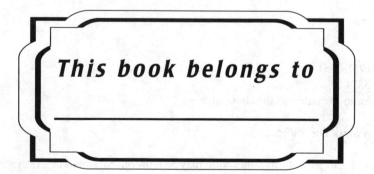

This book belongs to

_____

Electricity:
Science, Energy, Inventions

ISBN 1-888306-06-8

Copyright © 1997 by Amanda Bennett

Published by:
Homeschool Press
255 S. Bridge St.
P.O. Box 254
Elkton, MD 21922-0254
Tel. (410) 392-5554
Send requests for information to the above address.

Cover design by Mark Dinsmore.

Printed in the United States of America.

This book is dedicated to my brother and sisters.
Thanks for childhood years of static electricity
experiments–with socks "scrubbed" across the carpet
so we could zap the doorknobs, as well as each other!

How To Use This Guide

Welcome to the world of unit studies! They present a wonderful method of learning for all ages and it is a great pleasure to share this unit study with you. This guide has been developed and written to provide a basic framework for the study, along with plenty of ideas and resources to help round out the learning adventure.

TO BEGIN: The **Outline** is the study "skeleton", providing an overall view of the subject and important subtopics. It can be your starting point—read through it and familiarize yourself with the content. It is great for charting your course over the next few weeks (or developing lesson plans). Please understand that you do not necessarily have to proceed through the outline in order. I personally focus on the areas that our children are interested in first—giving them "ownership" of the study. By beginning with their interest areas, it gives us the opportunity to further develop these interests while stretching into other areas of the outline as they increase their topic knowledge.

By working on a unit study for five or six weeks at a time, you can catch the children's attention and hold it for valuable learning. I try to wrap up each unit study in five or six weeks, whether or not we have "completed" the unit outline. The areas of the outline that we did not yet cover may be covered the next time we delve into the unit study topic (in a few months or perhaps next year). These guides are **non-consumable**—you can use them over and over again, covering new areas of interest as you review the previous things learned in the process.

The **Reading** and **Reference Lists** are lists of resources that feed right into various areas of the **Outline**. The books are listed with grade level recommendations and all the information that you need to locate them in the library or from your favorite book retailer. You can also order them through the national Inter-Library Loan System (I.L.L.)—check with the reference librarian at your local library.

There are several other components that also support the unit study. They include:

The **Spelling and Vocabulary Lists** identify words that apply directly to the unit study, and are broken down into both Upper and Lower Levels for use with several ages.

The **Suggested Software, Games and Videos List** includes games, software and videos that make the learning fun, while reinforcing some of the basic concepts studied.

The **Activities and Field Trip List** include specific activity materials and field trip ideas that can be used with this unit to give some hands-on learning experience.

The **Internet Resources List** identifies stops on the information super-highway that will open up brand new learning opportunities.

The author and the publisher care about you and your family. While not all of the materials recommended in this guide are written from a Christian perspective, they all have great educational value. Please use caution when using any materials. It's important to take the time to review books, games, and internet sites before your children use them to make sure they meet your family's expectations.

As you can see, all of these sections have been included to help you build your unit study into a fun and fruitful learning adventure. Unit studies provide an excellent learning tool and give the students lifelong memories about the topic and the study.

The left-hand pages of this book have been filled with new and exciting things which will help bring this study to life for your family.

"Have fun &
Enjoy the Adventure!"

Table of Contents

Introduction

The flick of a switch and light floods the room. Set the microwave timer for one minute and a plate of leftovers is hot and ready to eat. The whirring of the washing machine, the beeping of a computer as it is turned on, the ringing telephone and the hum of the refrigerator—we depend on electricity for so many things in our daily lives! It is hard to imagine what life was like before the advent of electricity in our homes. We so often take it for granted until the power goes off or the light bulb burns out. The use of electricity has absolutely changed the way we live our lives and spend our time, making things easier and simpler while improving our comfort, our health and the quality of our lives.

God Himself has put electricity to work in the world around us. From the lightning in a storm to the flashing of signals through our nervous system, you can see this mysterious power at work everywhere in nature. People have not "discovered" or invented electricity—it has been here since creation! We have just been studying its mysteries, trying to understand its principles and learning to make it work for us in our daily lives.

The story of electricity is just that—a story of curiosity, intuition, exploration, and persistence. You can read about the ideas of Count Volta, the experiments of Benjamin Franklin, the electrification of rural America and the trials and successes of Thomas Edison. The people are just as interesting as the topic itself! Edison was home schooled by his mother, traveled as a telegraph operator during the Civil War and obtained over 1,000 patents in his lifetime!

As intimidating as a study of electricity might be at first glance, it is an unopened gift to be enjoyed by our children. The history and people involved are fascinating, but the properties, experiments, and unlimited ideas can absolutely captivate their minds. When students learn the basics of electricity and begin their experiments, they also learn to apply all of this new knowledge to their own ideas, like developing an alarm for their bedroom, set to go off when siblings enter. When encouraged, their ideas and questions can occupy many afternoons of tinkering, designing, and testing just to see what they can come up with and how they can make it work.

This study has been written to help us all learn about electricity in a very simple and engaging way. We can enjoy its history and the fun of experimentation, as well as how it affects our lives. This study covers areas such as:

- The history of electricity
- The inventors and scientists
- The properties of electricity
- Safety concerns with electricity
- Electricity in our daily lives

It is my hope that this study helps you and your students on the road to scientific adventure as well as on a wonderful journey back through history. Two simple things to remember—NEVER experiment with electricity without adult supervision and NEVER use a wall outlet as an electrical source for an experiment. There now—with that said, go on, pull out a fresh pack of batteries and a simple experiment book and enjoy yourself! The wires, lights and simplicity of electricity can awaken a new world of interests in your children, as well as teaching some important concepts along the way. Bring along all of the awe and wonder you can find as we begin a journey into a fun and lively topic—and as always, Enjoy the Adventure!

SIGN OF THE TIMES

1500-1550 A.D.

Here are some interesting events to include on YOUR time-line! Then, add on the important dates and events that you learn from your study of Electricity from this era . . .

1501 —Michelangelo begins the statue of David
1509 —Peter Heinlein invents the watch in Nuremberg
1513 —Ponce de Leon discovers Florida
1518 —Forks are used for the first time
1519 —Magellan leaves Europe to sail around the world
1529 —Martin Luther publishes <u>Away in a Manger</u>
1543 —Copernicus proposes planets revolve around sun

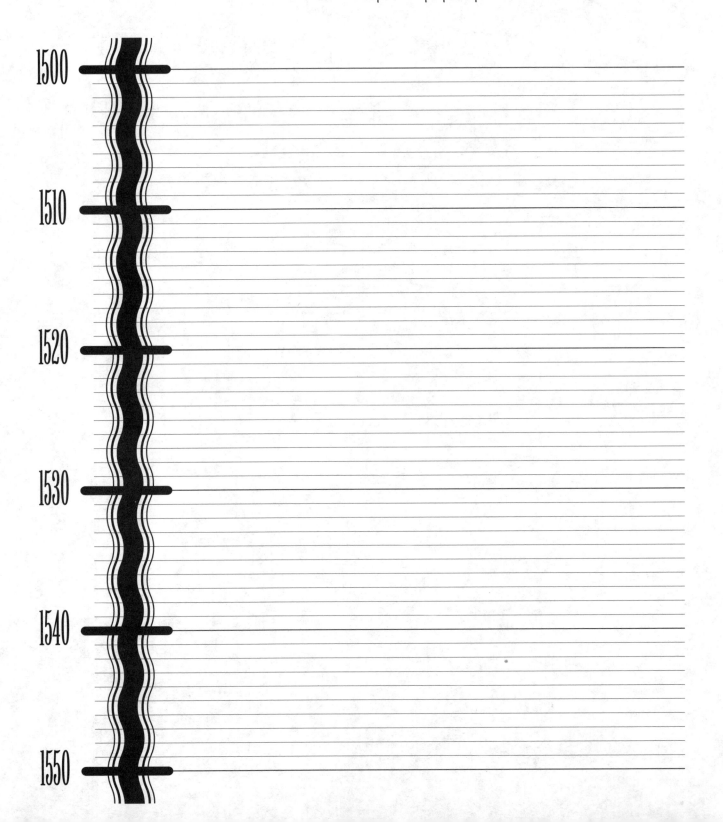

Unit Study Outline

I. Introduction

    A.   Origin and meaning of the word "electric"

    B.   Electricity is one of the many forms of energy created by God that surround us

    C.   Electricity in our daily lives
        1.   Our homes—electrical appliances such as lamps, air conditioning, stove, computer, etc.
        2.   In the air—static electricity and lightning
        3.   Our town—street lights, store signs, traffic lights, factories, stadium lights, etc.

    D.   Importance of electricity to people today
        1.   Increases the quality of our lives, by powering improvements like central air conditioning and heating, better lighting, better food preparation and preservation equipment, etc.
        2.   Improves our health standards, by providing better diagnostic equipment (X-ray, MRI, ultrasound) as well as better treatment options (lasers, endoscopic treatment, etc.)
        3.   Opens up the world around us, via television, multimedia sources, the Internet, etc., making people more aware of the needs, thoughts, and actions of others

II. Science of electricity

    A.   Types of energy (examples)
        1.   Electricity
        2.   Heat
        3.   Light
        4.   Sound

    B.   Two types of electricity
        1.   Static electricity—electricity at rest
        2.   Current electricity—electricity that travels through a complete circuit

Sign of the TIMES

1551-1600 A.D.

Here are some interesting events to include on YOUR time-line! Then, add on the important dates and events that you learn from your study of Electricity from this era . . .

1560 —Puritanism begins in England
1565 —Pencils are first manufactured in England
1578 —The catacombs of Rome are discovered
1582 —The Gregorian calendar is put into use
1590 —The compound microscope is invented
1596 —Galileo invents the thermometer

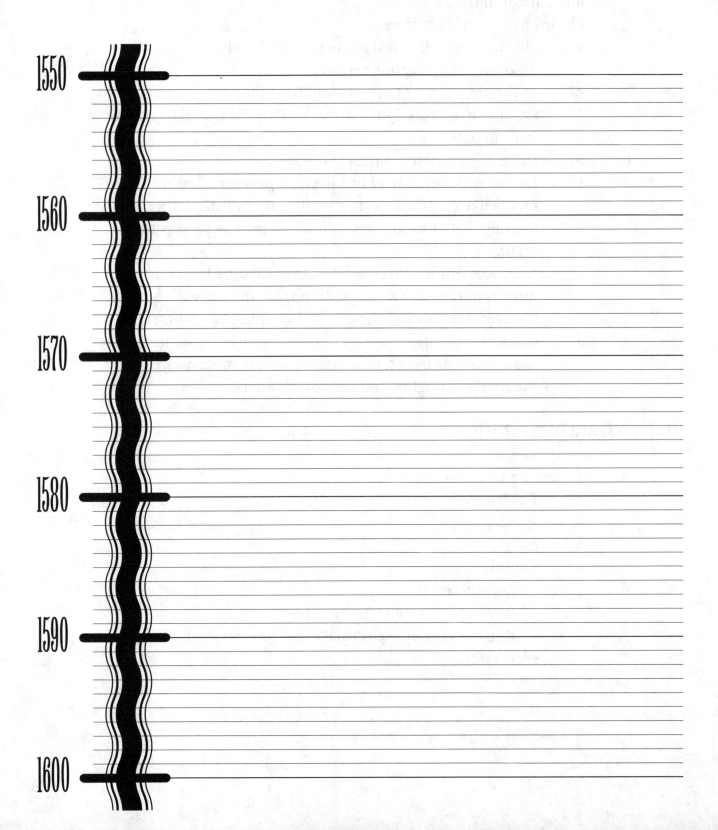

        a.   Alternating Current (AC)—where the electron flow changes direction constantly

        b.   Direct Current (DC)—where the electron flow is only in one constant direction

C.   Electricity and atoms

    1.   The three main parts of an atom

        a.   Proton—positively charged particle

        b.   Electron—negatively charged particle

        c.   Neutron—neutral particle (no charge)

    2.   Electricity is the flow of electrons from atom to atom

        a.   Opposite charges attract

        b.   Like charges repel

D.   Electrical flow through different materials

    1.   Conductors

        a.   Materials that allow a fast flow of electrons

        b.   Examples include gold, silver, copper

    2.   Insulators (nonconductors)

        a.   Materials that resist or impede the flow of electrons

        b.   Examples include wood, rubber, plastic

E.   Circuits

    1.   Defined as being a continuous or "closed" path for the flow of electricity

    2.   Types of circuits

        a.   Parallel—more than one path for the current to follow

        b.   Series—only one path for the current to follow

    3.   Common components of a circuit (conductors)

        a.   Wiring

        b.   Power supply (example: battery)

        c.   Switch

        d.   Example of miscellaneous components—vary by the application of the circuit

             (1)   Light bulb

             (2)   Buzzer

             (3)   Motor

             (4)   Resistor

             (5)   Capacitor

SIGN OF THE TIMES

1601–1650 A.D.

Here are some interesting events to include on YOUR time-line! Then, add on the important dates and events that you learn from your study of Electricity from this era . . .

1608 —The telescope is invented by Lippershey
1615 —Galileo faces the Inquisition
1620 —Pilgrims leave Plymouth, England and
        arrive in Massachusetts
1631 —The multiplication symbol (X) is introduced
1642 —The English Civil War begins

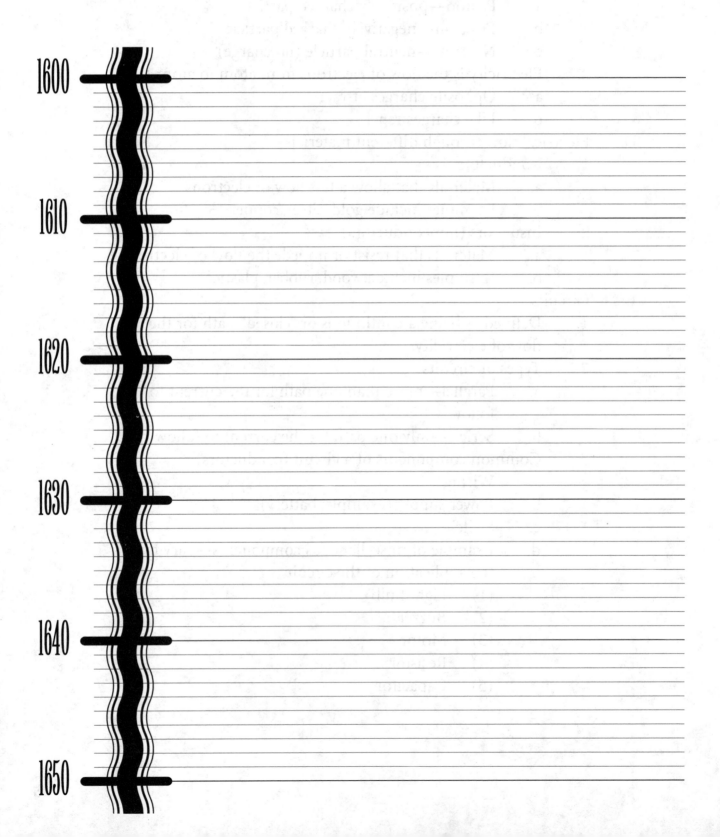

F.   Units of measurement associated with electricity
1.   Ampere (amp)—measures current
2.   Volt—measures potential energy
3.   Ohm—measures resistance to electrical flow
4.   Watt—measures power

III.   History, scientists, and inventors

A.   Ancient history of experimenting with electricity
1.   Thales of Miletus (625 - 547 B.C.)
a.   Greek mathematician and philosopher
b.   Discovered that amber stone (fossilized tree resin) attracted lightweight objects after being rubbed
2.   Pliny the Roman (First century A.D.)
a.   Statesman and writer
b.   Studying Thales' findings, Pliny performed some experiments of his own
c.   Wrote about Thales' work and his own experiments
B.   Renewed interest in electricity, 17th-18th centuries
1.   William Gilbert (1544-1603)
a.   Physician to Queen Elizabeth I of England
b.   In 1600, he wrote about his experiments with magnets and the attraction of various materials after being rubbed
c.   Named the materials that attract other materials "electrics" after the Greek word for amber, elektron
d.   Explained how the earth itself acts as magnet, offering an explanation of how the compass works
2.   Otto Von Guericke (1602-1686)
a.   German mayor of the town of Magdeburg
b.   Invented the first friction machine, used to generate electricity
c.   Constructed his friction machine from a ball made of sulfur, rubbing the ball with a cloth produced sparks of electricity

SIGN OF THE TIMES

1651–1700 A.D.

Here are some interesting events to include on YOUR time-line! Then, add on the important dates and events that you learn from your study of Electricity from this era . . .

1664 —Springs are designed and placed on travel coaches
1668 —Isaac Newton builds the first reflecting telescope
1675 —King Philip's War begins in New England
1679 —The pressure cooker is invented
1696 —The first practical steam engine is invented

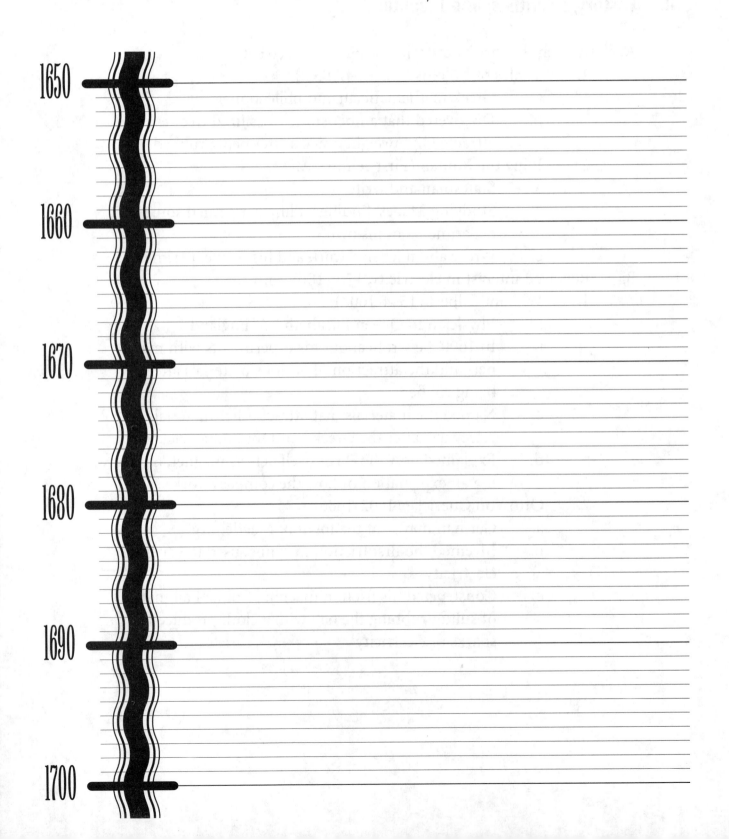

1650

1660

1670

1680

1690

1700

3. Stephen Gray (1666-1736)
    a. An English scientist, experimented in the 1720s with the movement of electricity, and found that electricity could move from one object to another
    b. Discovered that electricity moves easily through some materials, which he called conductors
    c. Materials that electricity does not flow through (silk, glass, etc.) were called non-conductors, which led to their use as insulators to control the flow of electricity
4. Charles Francis DuFay (1698-1739)
    a. French scientist, experimented in 1733 with electricity and found that there were two different kinds of electricity
        (1) Static electricity, which is generated from rubbing two objects together
        (2) Current electricity, where electricity actually moves through a material (electron flow)
5. In 1746, the Leyden jar was developed by scientists at the University of Leyden in the Netherlands, and used to store electrical charge
6. Benjamin Franklin (1706-1790)
    a. American scientist, inventor, and statesman
    b. Experimented with the behavior of electricity, discovering and naming objects as positively or negatively charged
    c. Proved that lightning was actually an electric spark with his famous kite experiment in 1752
    d. Invented the lightning rod for buildings, to prevent their destruction from lightning
7. Luigi Galvani (1737-1798)
    a. Italian physician and professor
    b. Discovered in 1771 that frog legs used for an experiment would twitch when they came in contact with two metals, calling this electricity "animal electricity"

Sign of the Times

1701–1750 A.D.

Here are some interesting events to include on YOUR time-line! Then, add on the important dates and events that you learn from your study of Electricity from this era . . .

1701 —Pennsylvania adopts its own Charter of Liberties
1705 —The ship's "wheel" is put in use, replacing the tiller
1711 —The tuning fork is invented
1727 —The first railroad bridge is built
1732 —Poor Richard's Almanac is published by Ben Franklin
1742 —Ben Franklin invents the Franklin stove

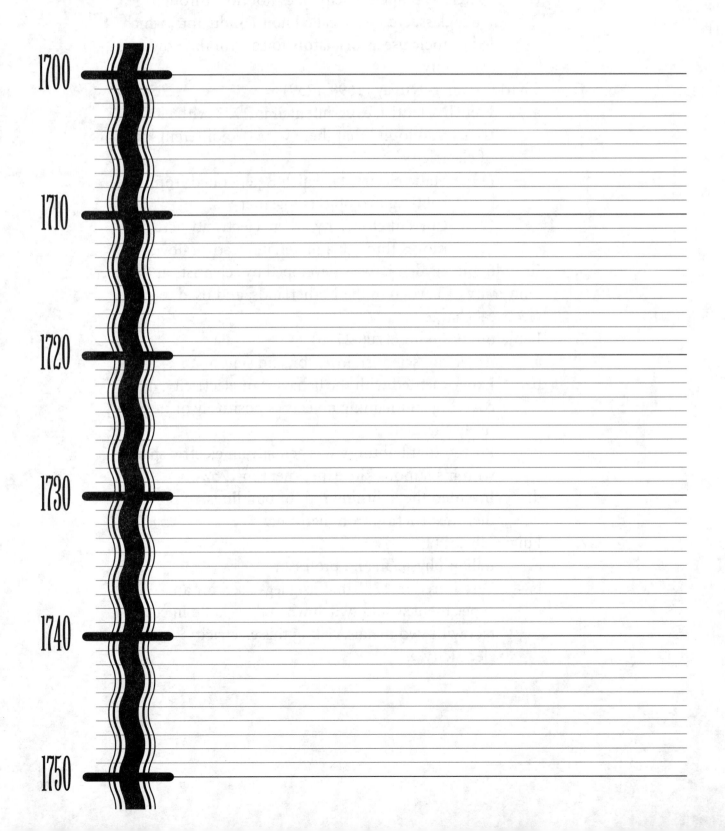

1700

1710

1720

1730

1740

1750

8. Alessandro Volta (1745-1827)
   a. Italian history professor, disagreed with Galvani's concept of "animal electricity"
   b. Worked with metals to develop the "voltaic pile", which was composed of various metal discs as well as other materials
   c. The voltaic pile produced a continuous electrical flow—the first battery
   d. The unit of measurement of electric potential, volt, is named after Volta

C. Discoveries in the nineteenth century and beyond
   1. Hans Christian Oersted (1777-1851)
      a. Professor of physics, University of Denmark
      b. In 1820, he discovered that magnetism was produced by an electric current
      c. The operation of the electric motor is based on his theory of electromagnetism
   2. André-Marie Ampère (1775-1836)
      a. French scientist, experimented and discovered ways to mathematically explain the magnetic fields generated by electric current
      b. The unit of measurement of electrical current is called the amp, named in his honor
   3. Georg Simon Ohm (1787-1854)
      a. German scientist, discovered that all materials have some form of resistance to electrical current—called this the "resistance" of the material, later to be measured in units of ohms, named in his honor
      b. Developed the mathematical relationship between resistance and current and electrical potential, called Ohm's law
   4. Michael Faraday (1791-1867)
      a. English scientist, called the "Father of Electricity" because of his discovery of the basics of electromagnetic induction, the principle of the electric motor
      b. Invented the modern electric generator (dynamo)

SIGN OF THE TIMES

1751–1800 A.D.

Here are some interesting events to include on YOUR time-line! Then, add on the important dates and events that you learn from your study of Electricity from this era . . .

1752 —Ben Franklin flies the kite in the storm
1765 —Stamp Act enacted; formation of the Sons of Liberty
1769 —James Watt obtains a patent for his steam engine
1770 —Beethoven is born
1771 —Galvani accidentally runs current through frog's leg
1773 —The Boston Tea Party is held
1783 —Treaty of Paris signed—War of Independence is over!
1788 —Our Constitution is ratified
1791 —Our Bill of Rights is ratified
1800 —Count Volta invents the electrical battery

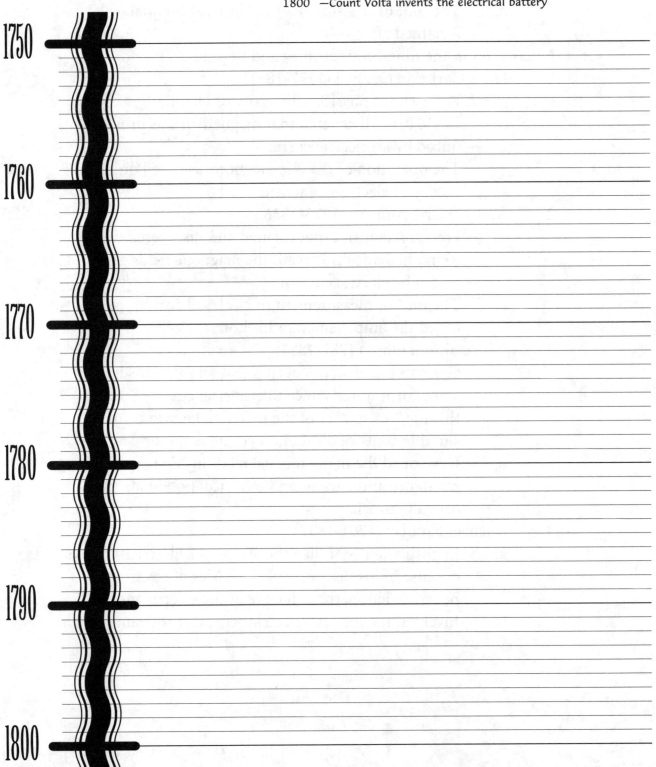

1750

1760

1770

1780

1790

1800

5. Samuel F. Morse (1791-1872)
   a. American artist, designed an electromagnetic telegraph and a code for sending messages on his telegraph (Morse code)
   b. In 1844, the first permanent telegraph transmitted Morse's famous first message, from Washington to Baltimore
6. Joseph Henry (1797 - 1878)
   a. American scientist, noted for his work with electro-magnetics
   b. Discovered the principle of self-induction, leading the way for the development of efficient electric motors
7. Alexander Graham Bell (1847-1922)
   a. Born in Scotland, and later became a United States citizen, invented the first practical telephone
   b. Worked on the study of sound and educating the deaf
8. Thomas Edison (1847-1931)
   a. Brilliant American inventor of both electrical and mechanical apparatus
   b. Held over 1,000 patents in his lifetime
   c. Opened the first power station
   d. Some of his greatest inventions include the electric light bulb, the phonograph, and the kinetoscope
9. Nikola Tesla (1856 - 1943)
   a. Engineer and inventor, native of Croatia, later emigrated to America
   b. Remembered for his discovery of alternating current as well as his work with motors

IV. Applications of electricity

A. Residential
   1. Comfort
      a. Air conditioning and heating systems
      b. Lighting for night time comfort and tasks
      c. Ceiling fan
      d. Security system

Sign of the TIMES

1801–1850 A.D.

Here are some interesting events to include on YOUR time-line! Then, add on the important dates and events that you learn from your study of Electricity from this era . . .

1803 —The Louisiana Purchase is made for $15 million
1812 —America declares war on Britain
1816 —The stethoscope is invented
1825 —The Erie Canal is finally opened
1839 —Photography is invented
1841 —The saxophone is invented by Adolphe Sax
1846 —The planet Neptune is discovered
1848 —Karl Marx writes the Communist Manifesto

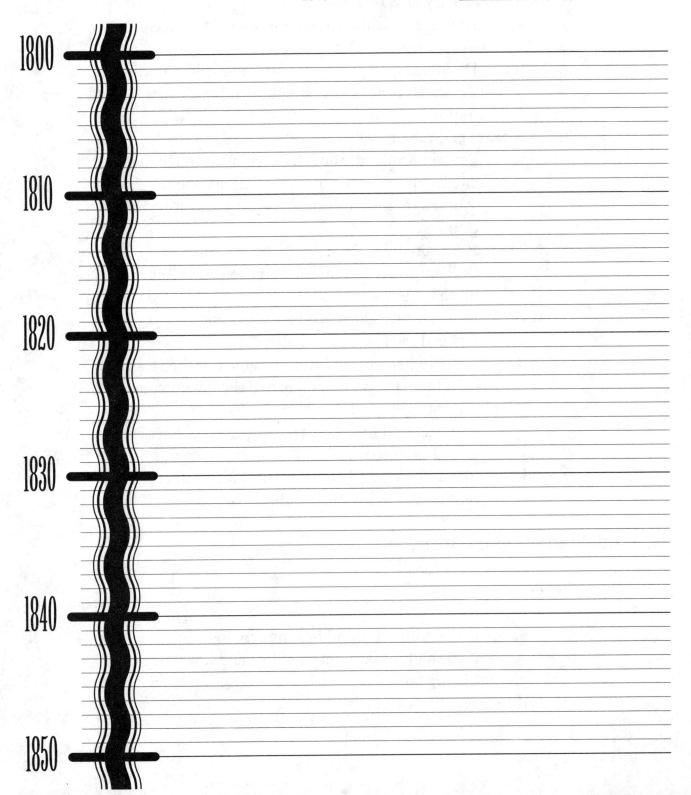

1800

1810

1820

1830

1840

1850

2. Household responsibilities
   a. Clothing care and construction
      (1) Sewing machine
      (2) Washing machine
      (3) Clothes dryer
      (4) Iron
   b. Food preparation
      (1) Refrigerator/freezer
      (2) Oven
      (3) Range
      (4) Microwave
      (5) Dishwasher
      (6) Small appliances like can openers, bread machines, toasters, etc.
   c. Home care and cleaning
      (1) Vacuum cleaner
      (2) Hot water heater
   d. Exterior home care and repair
      (1) Power tools
         (a) Drill
         (b) Saw
         (c) Compressor
      (2) Lawn care
         (a) Electric lawn mower
         (b) Hedge trimmer
         (c) Edger
3. Home business and recreation
   a. Computer equipment
   b. Fax machine
   c. Telephone equipment
   d. Television and VCR
B. Agricultural applications
   1. Food preparation equipment
   2. Animal care equipment
   3. Crop care equipment
C. Commercial applications (stores, banks, restaurants, etc.)
   1. Air conditioning and heating systems
   2. Food preparation equipment

Sign of the TIMES

1851–1900 A.D.

Here are some interesting events to include on YOUR time-line! Then, add on the important dates and events that you learn from your study of Electricity from this era . . .

1851 —Isaac Singer invents the sewing machine
1858 —The Lincoln-Douglas debates take place
1860 —South Carolina secedes from the Union
1864 —Louis Pasteur develops the process of pasteurization
1865 —The Civil War ends
1870 —Rockefeller founds the Standard Oil Company
1876 —Alexander Bell invents the telephone
1886 —The Statue of Liberty is dedicated
1889 —The Eiffel Tower is built

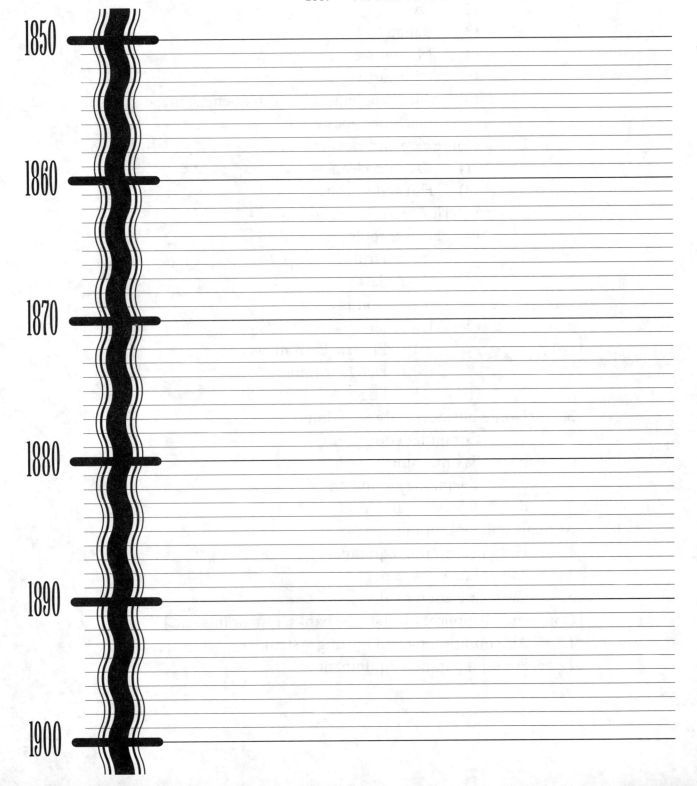

1850

1860

1870

1880

1890

1900

3. Business equipment
   a. Computer equipment
   b. Cash register
   c. Copy machine
   d. Fax machine
   e. Adding machine
4. Security system
D. Industrial applications (manufacturers)
   1. Production equipment
   2. Inventory management equipment
   3. Business equipment
   4. Comfort equipment
   5. Safety equipment

V. Impact of innovations and inventions that utilize electricity over the last 150 years

A. Economic impact
   1. Major change in industrial efficiency and productivity resulted from the use of electricity
   2. As industry changed and became more productive, economies grew and strengthened
   3. With the growing economies came more investment in industry, resulting in new interest in research and development to find new products
B. Worldwide changes
   1. Electricity is now produced around the world
   2. Products can now be manufactured in areas that previously would not have been considered
   3. Widespread use of these innovations has transformed the world in how we live, work, and play

VI. Electricity today—the production and transmission by utility companies

A. Electricity is now generated by major utility companies in several ways
   1. Fossil fuel plants (coal-fired)
   2. Nuclear plants

SIGN OF THE TIMES

1901–1950 A.D.

Here are some interesting events to include on YOUR time-line! Then, add on the important dates and events that you learn from your study of Electricity from this era . . .

1901 —Marconi sends the first transatlantic radio signal
1903 —The Wright brothers fly at Kitty Hawk, North Carolina
1908 —The Geiger Counter is invented
1915 —The British Lusitania is sunk by German submarines
1917 —America declares war on Germany (World War I)
1928 —Alexander Fleming isolates penicillin
1929 —The worldwide Depression begins
1932 —Franklin Roosevelt begins promoting "The New Deal"
1941 —Japanese bomb Pearl Harbor; U.S. declares war
1945 —World War II ends

1900

1910

1920

1930

1940

1950

3. Hydro-electric plants (using dams)
4. Alternative sources
   a. Solar equipment
   b. Wind turbines

B. Transmitting the electricity from the power plant to your home
   1. Electricity is transmitted through what is known as "the grid"—a network of cables connected throughout the country for the sole purpose of delivering electricity
   2. From the power plant to your home, the electricity will go through several sets of transformers to bring the power to residential requirements (110 Volts)
   3. At your home, the power will come through a fuse box or breaker box, delivering power to the various areas or zones of your home

VII. The future holds bright promise and many possibilities for new and improved inventions as well as applications of electricity in many areas of our lives

A. Medicine
   1. Better scanning methods
   2. Enhanced medication delivery systems (shots that won't hurt!)
   3. Improved methods of surgery

B. Electronics
   1. Increased efficiency
   2. Larger memory on smaller chips
   3. Faster operating systems that use less electricity

C. Communications
   1. Smaller and more affordable telephones using cellular or other satellite technology
   2. Improved satellites for communications
   3. Effective ways to utilize the Information Highway

D. Education
   1. Improved computers and software systems to increase educational options
   2. Virtual technology applications in education
   3. Online books and resources will become more readily available using smaller and more efficient satellite technology and cable systems

Sign of the TIMES

1951–2000 A.D.

Here are some interesting events to include on YOUR time-line! Then, add on the important dates and events that you learn from your study of Electricity from this era . . .

1950 —The Korean War begins
1954 —Jonas Salk develops a polio vaccine
1958 —NASA is established
1959 —Alaska and Hawaii become states
1961 —The Berlin Wall is erected
1962 —John Glenn is the first American in space
1967 —The first human heart is transplanted
1969 —Neil Armstrong is the first man to walk on the moon
1973 —U.S. troops withdraw from Vietnam
1989 —The Berlin Wall is dismantled

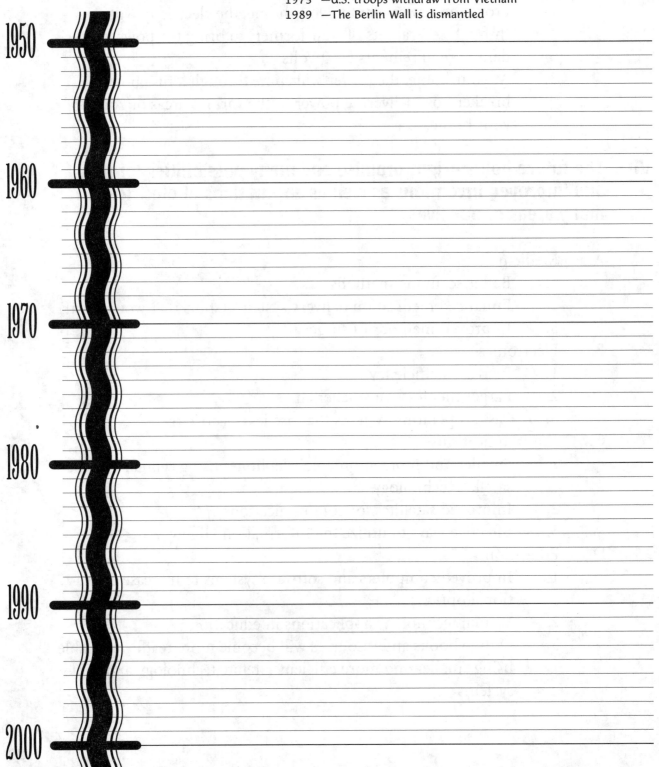

1950

1960

1970

1980

1990

2000

E. Transportation
   1. Improved electric cars
   2. New ways to utilize electricity in mass transit
   3. Improved systems and methods of propulsion and energy for the space program
F. Government
   1. Improved voting equipment and counting systems
   2. With the vast media information becoming more quickly and readily available, accountability of our elected officials should improve
   3. Better use of computers and new scanning and reporting systems will keep the public much more aware of government operations

SHOW what you KNOW!

Check out these words from the **Lower Level** Spelling/Vocabulary List, and SHOW what you KNOW!

wire

magnet

radio

amp

battery

Morse code

light

lamp

socket

switch

motor

outlet

safe

prong

power

danger

on/off light bulb socket open/closed battery electric source circuit wind outlet plug safe radio wave careful power danger motor lamp morse code volt flash magnet wire watt switch push pull buzzer

Spelling and Vocabulary
Lower Level

hot	radio
rub	amp
shock	battery
touch	Morse
feel	code
run	bell
stop	ring
wire	light
go	bulb
key	lamp
kite	socket
care	switch
fire	complete
path	try
clip	motor
coil	scale
line	home
pole	plug
turn	outlet
see	safe
on	open
off	close
chip	prong
look	travel
magnet	power
crank	plant
hour	tower
tube	overhead
get	danger
flow	avoid

SHOW what you KNOW!

Check out these words from the **Upper Level** Spelling/Vocabulary List, and SHOW what you KNOW!

alternating current

capacitor

charge

conductor

current

direct current

electricity

electrode

electron

energy

insulator

kilowatt

magnetic

power

resistor

voltage

magnet filament motor series parallel power circuit design kilowatt ohm capacitor shock electric conductor insulator current meter switch light electron experiment battery hydro coal fossil solar charge

Spelling and Vocabulary
Upper Level

electricity

current

amber

static

charge

lightning

battery

attraction

magnetic

quantify

conductor

electrode

insulator

armature

motor

resistor

capacitor

condenser

aluminum

relay

scientific

observation

experiment

repel

electron

induction

storing

circuit

switch

device

ohm

parallel

series

voltmeter

filament

resistance

application

schematic

diagram

continuity

generator

force

electrostatic

alternating current

direct current

measurement

meter

telegraph

voltage

prediction

kilowatt

utilities

energy

power

transmission

turbine

transformer

substation

grid

utility

Who?

What?

When?

Where?

Why?

Electricity • Electricity • Electricity • Electricity • Electricity • Electricity • Electricity • Electricity • Electricity • Electricity

Writing Ideas

This section has been included to suggest some ideas on how to use writing in a unit study. Choose one or two and watch what happens!

1. Have the students write to your local or regional utility company, requesting information on electricity, safety, conservation and other information that they might be able to provide free of charge. Remind the students to mention their age and/or grade level to be certain that the material will be age-appropriate.

2. A fun and very informative writing adventure is to have the children keep an energy journal for a few weeks. Have them enter all of the times they use or rely upon electricity for their daily lives. They can carry around a small wire notebook and make an entry for each event where they use electricity. Some entry examples might include cooking, lighting, alarm clock, heating, computer, etc.

3. After keeping the energy journal for a few weeks, have the students summarize their findings, describing the major uses as well as the level of importance of electricity in their lives. A child that lives in a rural home, where woodburning stoves and other alternative energy sources are common, might not place the same level of importance on electricity as a child in an urban home. When summarizing the findings in their report, have them also include a personal plan for cutting back on their own electrical consumption.

4. Energy conservation is very important, and is particularly emphasized in the use of electricity. Your local utility can provide literature for educational purposes that you might want to use with this unit. Use this literature along with other resources the students might find in the library, and have the students develop a written plan of energy conservation for your family's household.

PROJECT Ideas

Where we cannot invent, we may at least improve. —Colton

Invention is the talent of youth, as judgment is of age. —Swift

Things to try, places to see, papers to write, books to read, things to make . . .

Activity Ideas and Resources

Activities are a great way to reinforce the material that we learn. They provide important hands-on learning while allowing the student to have fun and be challenged at the same time. REMEMBER, please, that all experiments and activities dealing with electricity should have ADULT supervision, and make sure that the children are aware of this. Here are some activity resources to consider using with this unit:

The Thomas Edison Book of Easy and Incredible Experiments: Activities, Projects and Science Fun for All Ages, by The Thomas Alva Edison Foundation. Grades 4 - 11. Published by John Wiley & Sons, Inc., 1 Wiley Drive, Somerset, NJ 08875. (800) 225-5945.

Electricity Video Science Lab, includes an instructional video and lab worksheets, as well as the required electrical parts to perform the various lab experiments. Grades 4 - 7. Available from Farm Country General, Rt. 1 Box 63, Metamora, IL 61548. (800) 551-FARM.

The Benjamin Franklin Book of Easy & Incredible Experiments: Activities, Projects, and Science Fun, a Franklin Institute Science Museum Book. Grades 3 - 8. Published by John Wiley & Sons, Inc., 1 Wiley Drive, Somerset, NJ 08875. (800) 225-5945.

Benjamin Franklin: Scientist & Inventor, Activity Book, from Living History Productions, Inc. Grades 2 - 6. Can be used independently or as a supplement to the video "Benjamin Franklin", from Living History Productions. Both are available from Family Christian Academy, 487 Myatt Drive, Madison, TN. (800) 788-0840 ext. 3009.

Fun With Electronics, Jr., by Luann Colombo and Conn McQuinn, for grades K - 4 and **Fun With Electronics**, by Conn McQuinn for grades 5 and up. Both of these kits come with electronic parts and an instruction booklet to give instructions for some fun projects. Published by Andrews and McMeel, Universal Press Syndicate Co. Available from Tobin's Lab, P.O. Box 6503, Glendale, AZ 85312-6503. (800) 522-4776.

Electric Motor Kit, a Mini Science Lab Kit. Provides the parts and instructions required to build a small electric motor. Available from Tobin's Lab, P.O. Box 6503, Glendale, AZ 85312-6503. (800) 522-4776.

Resources @HOME

The first sure symptom of a mind in health, is rest of heart, and pleasure felt at home.

—Young

To Adam paradise was home. To the good among his descendants, home is paradise.

—Hare

Tools, Books, Toys, Materials, Hobbies, Internet Sites...

Electricity • Electricity • Electricity • Electricity • Electricity • Electricity • Electricity • Electricity • Electricity • Electricity • Electricity • Electricity • Electricity • Electricity • Electricity • Electricity • Electricity • Electricity • Electricity

Internet Resources

Here are some interesting sites on the Internet that you might want to visit while studying this unit. Please keep in mind that these pages, like all web pages, change from time to time. I recommend that you visit each site first, before the children do, to view the content and make sure that it meets with your expectations. Also, use the **Subject Key Words** for search topics on Internet search engines, to find the latest additions that might pertain to this topic.

Experiments and Ideas:

Electric Battery Experiments with the Energizer® Bunny
http://www.energizer.com/Science_Projects/shtr_main.html

Build a Galvanometer
http://schoolnet2.carleton.ca/english/math_sci/phys/electric-club/page11.html

Electricity and Magnetism
http://www.mip.berkeley.edu/physics/bookddx.html

Energy Encounter
http://www.flp.com/fplpages/encountr.htm

You Can With Beakman
http://www.nbn.com/youcan/

Bill Nye the Science Guy
http://www.nyelabs.kcts.org/

Inventions:

Invention Dimension
http://web.mit.edu/invent/

Invention, Design, and Discovery
http://jefferson.village.virginis.edu/-meg3c/id/id_home.html

What Makes an Inventor Successful?
http://mustang.coled.umn.edu/inventing/inventing.html

Tomorrow's Inventions
http://sin.fi.edu/franklin/inventor/inv_now.html

PLAN & Investigate

A contemplation of God's works, a generous concern for the good of mankind, and the unfeigned exercise of humility—these only, denominate men great and glorious.

—Addison

Plan your work, then work your plan!

_____
_____
_____
_____
_____
_____
_____
_____
_____
_____
_____
_____
_____
_____
_____
_____
_____
_____
_____
_____
_____
_____

Electricity • Electricity • Electricity • Electricity • Electricity • Electricity • Electricity

Inventors and Scientists:

Alexander Graham Bell's Path to the Telephone
http://jefferson.village.virginia.edu/albell/homepage.html

Bell's Telephone
http://sin.fi.edu/franklin/inventor/bell.html

Alessandro Volta's Shocking Discovery
http://www.adventure.com/library/encyclopedia/ka/rfivolta.html

Nikola Tesla
http://www.invent.org/book-text/102.html

Benjamin Franklin, Inventor
http://sin.fi.edu/franklin/inventor/inventor.html

Electrified Ben
http://sin.fi.edu/franklin/scientst/faraday.html

Thomas Edison
http://sin.fi.edu/franklin/scientst/edison.html

Edison Lights the Night
http://www.adventure.com/library/encyclopedia/america/edison.html

Thomas Alva Edison
http://www.invent.org/book/book-text/38.html

Samuel Morse and Morse Code
http://www.vpds.wsu.edu/fair_95/gym/um122.html

Samuel Morse
http://www.acusd.edu/~ekarakis/Morse.html

Innovatons by Samuel Morse:
http://www.nmaa.si.edu/secrets/secretshtml/innovation.html

Inventure Place-The National Inventors Hall of Fame
http://www.invent.org/

Resources @HOME

The first sure symptom of a mind in health, is rest of heart, and pleasure felt at home.

—Young

To Adam paradise was home. To the good among his descendants, home is paradise.

—Hare

Tools, Books, Toys, Materials, Hobbies, Internet Sites...

_____
_____
_____
_____
_____
_____
_____
_____
_____
_____
_____
_____
_____
_____
_____
_____
_____
_____
_____
_____
_____
_____
_____
_____
_____

Electricity • Electricity • Electricity • Electricity • Electricity • Electricity • Electricity • Electricity

Museums and Science Centers:
Museum of Science & Industry
http://www.msichicago.org/

Sciencenter
http://edison.scictr.cornell.edu/

The Franklin Institute Science Museum
http://sin.di.edu/

Home Electricity:
Home Energy Information
http://www.energyoutlet.com

Electrical Safety
http://hammock.ifas.ufl.edu/txt/fairs/28552

Incandescent Lamps
http://www.interlectric.com/incand.html

Electric Utilities:
Electric Power Research Institute (EPRI)
http://www.epri.com/

Edison Electric Institute
http://www.eei.org

National Rural Electrification Co-op
http://www.nreca.org

U.S. Department of Energy
http://web.fie.com/fedix/doe.html

Utility Links
http://www.intr.net/pma/utillnks.html

Power by Niagara
http://www.niagaranet.com/niagara/power.html/

PLAN & Investigate

A contemplation of God's works, a generous concern for the good of mankind, and the unfeigned exercise of humility—these only, denominate men great and glorious.

—Addison

Plan your work, then work your plan!

Electricity • Electricity • Electricity • Electricity • Electricity • Electricity • Electricity • Electricity • Electricity • Electricity • Electricity

Electricity • Electricity • Electricity • Electricity • Electricity • Electricity • Electricity

Miscellaneous:

Thunderstorms and Lightning
gopher://esdim1.esdim.noaa.gov/00/NOAA_systems/education/thunder_and_
lightning

Historical Recordings
http://www.classical.net/~music/comp.lst/historic.html

Westinghouse
http://www.westinghouse.com

Westinghouse Science Talent Search
http://www.westinghouse.com/cmty/h_sci.htm

The History of Telephones
http://www.cybercomm.net/~chuck/phones.html

U.S. EPA Energy Star Programs and Products
http://www.epa.gov/docs/GCDOAR/EnergyStar.html

Famous Inventors

While there are many inventors that played key roles in the development of our use of electricity, there are several very important ones to investigate. Here are a few names—choose one and see what kind of interesting information you can find about this person! For example, did you know that Samuel Morse was at one time a portrait artist?

When and where was the inventor born? What kind of education did he receive? How did he get interested in inventing?

_____
_____
_____
_____
_____
_____
_____
_____
_____
_____
_____
_____
_____
_____
_____
_____
_____
_____
_____
_____
_____

Thomas Edison	Samuel Morse	Hans Christian Oersted
Benjamin Franklin	Alessandro Volta	Joseph Henry
James Watt	André-Marie Ampere	Alexander Bell
Luigi Galvani	Michael Faraday	Nikola Tesla
	George Ohm	

Job Opportunities

Here is a list of some of the jobs that involve electricity. There are others that I'm sure you will identify, but these are some of the main ones that we investigated during our unit study.

Electrical Engineer

Medical Equipment Designer

Mechanical Engineer

Appliance Repair Technician

Computer Science Engineer

Electronics Technician

Electronics Engineer

Power Plant Technician

Power Generation Engineer

Maintenance Electrician

Power Plant Operator

Electrician

Communications Engineer

Computer Hardware Designer

Communications Technician

For more information about these jobs or others that may be interesting, go to the reference librarian in the public library and ask for publications on careers. Some that we recommend are:

The Encyclopedia of Careers and Vocational Guidance, published by J. G. Ferguson Publishing Company, Chicago.

Occupational Outlook Handbook, published by the U.S. Department of Labor, Bureau of Labor Statistics. It presents detailed information on the 250 occupations that employ the vast majority of workers. It describes the nature of work, training and educational requirements, working conditions and earnings potential.

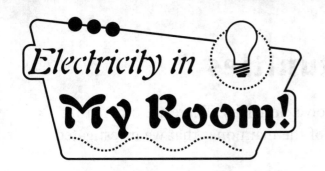

Electricity in My Room!

Draw a diagram of your room here. Mark the wall switches, electrical outlets, and light fixtures in your room. You can also indicate the locations of windows, doors, and furniture.

Suggestion: Take measurements for dimensions of larger objects, and use a ruler to create a scale diagram. Try this scale: 1/2 inch equals 1 foot.

Room Decorations

When working on a unit study, we try to decorate the room with items that relate to our current topic of interest. This allows the students to see the important information on a regular basis, as well as providing a place to view their own work. For electricity, consider some of the following ideas:

1. As you progress through the unit, have the students draw and diagram various components related to electricity on poster board for display. Consider items like a light bulb, a simple circuit and an experiment that they worked on to diagram on the posters.

2. Have the older students work on a schematic drawing of the electrical wiring within their room or the den, etc., by visually identifying outlets, lighting fixtures, switches and so on. This can be interesting—often, they have never given a second thought to the electrical circuitry in their own rooms.

3. Utility companies frequently have poster contests for students, based on themes like electricity safety and energy conservation. Contact your local utility company to find out if one is usually held in your area, and encourage your students to participate. Place the results of their efforts on your own walls when the competition is complete.

4. Once the child has learned the basics of electricity and some of the components, consider having him design his dream circuit for a display. Using poster board for the final version, let them display their own inventive ideas. Even though the circuit may not work in reality, the concepts demonstrated are usually very interesting, along the lines of burglar alarms in favorite toy boxes, door buzzers for their bedroom doors and on and on. There's a bit of God's creativity in all of these students!

Favorite
BIOGRAPHIES

A biography is a book that is written about someone's life. It provides a look "inside" a person's background, home life, education, and other areas that help us better understand the person that we are studying. There are many listed in this book to give you some ideas.

Who . . . what . . . where . . . when . . . why . . . Who . . . what . . . where . . . when . . . why . . . Who . . . what . . . where . . . when . . . why . . . Who . . . what

wh ere . . . when . . . why . . . Who . . . what . . . where . . . when . . . why . . . Who . . . what

List the biographies that you enjoyed here, along with the author and a description of the book!

1. _____
_____
_____
_____
_____

2. _____
_____
_____
_____
_____

3. _____
_____
_____
_____

4. _____
_____
_____
_____

Video Suggestions

While learning about electricity, there are many great television shows and videos available. Many of these can be obtained through your local library or video store.

Check with your local library and see what videos they have in their collection. We found some great ones about general electricity experiments, including some from television show scientists like Bill Nye the Science Guy and others. PBS has created several series and documentaries on the topic of electricity as well as inventors, and many libraries include these specials in their collections. Some libraries also loan out copies of videos made by local utility companies, on topics like safety issues of electricity, lightning, household safety, power plants, etc.

Look for videos or television documentaries on people that were key figures in the development of the use of electricity. Some of these include Thomas Edison, Benjamin Franklin, Alexander Graham Bell, Samuel Morse, etc. There are three videos from the Living History Production Series of videos of Animated Hero Classics that complement this unit. They are **Thomas Edison**, **Benjamin Franklin: Scientist and Inventor** and **Alexander Graham Bell**. This series is available through Family Christian Academy, 487 Myatt Drive, Madison, TN. (800) 788-0840.

Many videos are also available for older students on electricity as it applies to home improvement or remodeling projects. This type of video is also carried by libraries and video stores, as well as some home improvement stores. Topics include wiring in the home, small appliance repair, etc.

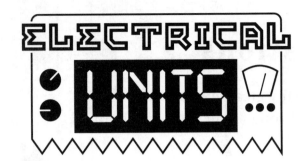

ELECTRICAL UNITS

When working on this study of electricity, see if there are other electrical units of measure that you can identify and list them here!

Electricity • Electricity • Electricity • Electricity • Electricity • Electricity • Electricity • Electricity • Electricity

Coulomb _____

Watt _____

Volt _____

Amp _____

Joule _____

Ohm _____

Farad _____

_____

_____

_____

_____

_____

_____

_____

_____

_____

_____

_____

Electricity • Electricity • Electricity • Electricity • Electricity • Electricity • Electricity

Games and Software

Games are a great way to reinforce the material that we learn. We have fun, while reviewing important information and concepts around the kitchen table or on the computer. The software listed here is just a small sample of what is available. With the writing of this book, there are several new games and software packages in development for release in the near future, and many sound very exciting! Check around at your local toy and software stores to find out the latest introductions.

Games:

AC/DC Game, a card game that helps teach the basic principles of electricity. Available from Tobin's Lab, P.O. Box 6503, Glendale, AZ 85312-6503. (800) 522-4776.

Inventors Card Game, by U.S. Games Systems. Available from Tobin's Lab, P.O. Box 6503, Glendale, AZ 85312-6503. (800) 522-4776.

Software:

Thinkin' Things, Collection III, problem-solving skills software for ages 7 and up. Published by Edmark Corporation, P.O. Box 97021, Redmond, WA 98073-9721. (800) 320-8379.

InventorLabs, from Houghton Mifflin Interactive, 120 Beacon Street, Somerville, MA 02143. (800) 829-7962. Software for grades 5 and up, taking the user to the labs of three great inventors–Thomas Edison, Alexander Graham Bell, and James Watt.

EKI Home School Electronics Discovery Software and Lab, from Electronic Kits International, 178 South State St., Orem, UT 84058. (800) 453-1708. Grades 6 - 12.

Your UTILITY COMPANY

As you will learn in this study, the utility company is responsible for providing electricity to your home. There are many different ways that utility companies can produce the electricity. Do your own investigation—what types of power plants does YOUR utility company use? Call them to find out. Then sketch the locations of the various facilities on a map of your area below. You may also wish to see if they provide tours for field trips!

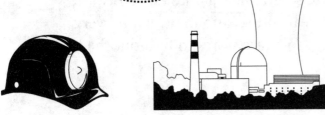

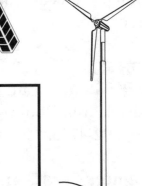

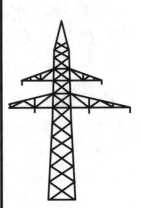

Field Trip Ideas

There are so many field trips that can be enjoyed while learning about electricity, that it is hard to list all of the ones that you might want to consider. Please use this list to get started planning some field trips, then let your imagination identify others that are in your area. Don't forget to take along your camera to capture some of the sights of the surroundings as well as the children! Also, write to the places that you visit and thank them for their time, or have one of the older children do the note-writing.

1. Arrange a visit to your local or regional utility company by contacting their public relations or customer service department. They frequently offer tours of their power plant facilities, educational resource centers and other special areas of interest to students (dams, co-generation facilities, etc.).

2. If there happens to be a manufacturing plant in your area that produces electrical appliances, motors, components, computers, etc., contact their public relations department to arrange a plant tour for your students. This can be very interesting, as well as educational, for students and parents alike. One of my favorite tours was of a small appliance manufacturing facility, where you could observe the installation of all of the electrical components along an assembly line, and watch the testing of the finished product for performance as it neared the end of the production line.

3. Since electricity can be very dangerous, many fire departments, insurance companies, etc., offer home safety courses to the public from time to time. Sign up the students for one of these classes, or better yet, take the class as a family. This can be a lifesaver for families, and is well worth the time invested for all members to attend the classes. Just having a small child that can dial 911 in an emergency has saved many lives!

4. An interesting trip to make during this study is to a neighborhood appliance repair shop. Your students can have the opportunity to see many different kinds of electrical appliances all taken apart, being repaired and then reassembled.

Our delight in any particular study, art, or science rises and improves in proportion to the application which we bestow upon it. Thus, what was at first an exercise becomes at length an entertainment.

—Addison

List some questions for which you would like to research the answers.

Who . . . what . . . where . . when . . . why . . . Who . . . what . . . Where . . . when . . . why . . . Who . . . what . . . where . . . when . . . why . . . Who . . . what

Subject Word List

This list of **SUBJECT** search words has been included to help you with this unit study. To find material about electricity, go to the card catalog or computerized holdings catalog in your library and look up:

General Words

alternating current (AC)
battery
circuit
direct current (DC)
electric appliance
electric
electric light
electric meter
electric motor
electrical engineering
electrical repair
electrical
electrician
electricity
electromagnetic
electronic
energy
energy conservation
experiment
fluorescent light
fossil power
generator
household safety
hydro power
incandescent light
light
magnet
nuclear power

power
power plant
radio
semiconductor
solar battery
solar power
telegraph
telephone
television
utility

People

Thomas Edison
Benjamin Franklin
Alessandro Volta
Alexander Graham Bell
Samuel Morse
Georg Ohm
Michael Faraday
Henry Cavendish
Luigi Galvani
Charles de Coulomb
Heinrich Hertz
James Watt
Thales of Miletus
Nikola Tesla
André-Marie Ampère
Hans Christian Oersted

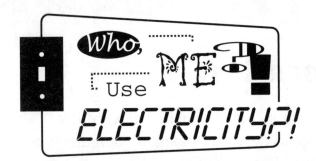

Without being aware of it, we ALL use electricity—almost all of the time! We use it for light, air conditioning, heating, television, computers, stereo, cooking, washing our clothes, and even for alarm clocks! Here's your chance to see just how much YOU use electricity.

Use this page to keep track of each and every time you use something electrical in a 24-hour period. Don't forget the refrigerator! ☺

Time	Item	How long?

Run out of room? We thought you might! Photocopy this page, or use a separate sheet of paper.

Trivia Questions

These questions have been included for fun, as well as to reinforce some of the material that you might read during this study. Enjoy the search for answers, and then compare them with the answers that we found which are located in the next section!

1. What inventor is known as the "Father of Electricity"?

2. What American inventor/manufacturer successfully developed an electric motor-driven sewing machine?

3. What type of electrical generator is run by water power?

4. Who invented the lightning rod?

5. The "watt" is a unit measure of power, named in honor of James Watt, a Scottish inventor. What invention is he most remembered for?

6. The inventor of the telephone, Alexander Graham Bell, was also known for his work in the education of people with what kind of disability?

7. Before inventing the telegraph, Samuel Morse pursued what kind of career?

8. What was the first message sent by Samuel Morse on his first permanent telegraph?

9. Who discovered alternating current (AC)?

10. What inventor was the founder of the National Geographic Society?

Trivia Answers

1. What inventor is known as the "Father of Electricity"?
Michael Faraday

2. What American inventor/manufacturer successfully developed an electric motor-driven sewing machine?
Isaac Singer

3. What type of electrical generator is run by water power?
Hydroelectric

4. Who invented the lightning rod?
Benjamin Franklin

5. The "watt" is a unit measure of power, named in honor of James Watt, a Scottish inventor. What invention is he most remembered for?
The first efficient steam engine

6. The inventor of the telephone, Alexander Graham Bell, was also known for his work in the education of people with what kind of disability?
Deafness

7. Before inventing the telegraph, Samuel Morse pursued what kind of career?
Portrait painter

8. What was the first message sent by Samuel Morse on his first permanent telegraph?
"What hath God wrought!"

9. Who discovered alternating current (AC)?
Nikola Tesla

10. What inventor was the founder of the National Geographic Society?
Alexander Graham Bell

Reference Resources
History

Electricity, by Steve Parker. (Eyewitness Science Series). Grades 6 - 12. Published by Dorling Kindersley, DK Family Learning, 7800 Southland Blvd, Suite 200, Orlando, FL 32809 (800) 352-6651, distributed by Houghton Mifflin, Industrial/Trade Division, 181 Ballardvale Rd., Wilmington, MA 01887 (800) 225-3362.

The Lightbulb & How It Changed the World, by Michael Pollard. (History & Invention Series). Published by Facts on File, 11 Penn Plaza, 15th Floor, New York, NY 10001. (800) 322-8755.

What It Was Like Before Electricity, by Paul Bennett. (Read All About It Series). Grades 1 - 8. Published by Raintree Steck-Vaughn, P.O. Box 26015, Austin, TX 78755. (800) 531-5015.

The Light Bulb: Inventions That Changed Our Lives, by Sharon Cosner. Grades 5 and up. Published by Walker Publishing Co., 435 Hudson Street, New York, NY 10014. (800) 289-2553.

How Did We Find Out About Electricity?, by Isaac Asimov. (How Did We Find Out About... Series). Grades 5 - 8. Published by Walker Publishing Co., 435 Hudson Street, New York, NY 10014. (800) 289-2553.

Experiments & Observations on Electricity, by Benjamin Franklin. (Notable American Author Series). Published by Reprint Services Corporation, P.O. Box 890820, Temecula, CA 92589. (909) 699-5731.

Electricity: From Faraday to Solar Generators, by Martin Gutnik. (First Book Series). Grades 4 - 9. Published by Franklin Watts, 5450 Cumberland Avenue, Chicago, IL 60656. (800) 672-6672.

Electronics, by Roger Bridgman. (Eyewitness Science Series). Grades 5 - 12. Published by Dorling Kindersley, DK Family Learning, 7800 Southland Blvd, Suite 200, Orlando, FL 32809 (800) 352-6651, distributed by Houghton Mifflin, 181 Ballardvale Rd., Wilmington, MA 01887. (800) 225-3362.

Simple INVENTIONS

Many of the simple items we use today have some fascinating histories. Look around the house and choose a simple item from your daily routine. It might be a paper clip, toothbrush, Velcro® fasteners on your tennis shoes, a pencil, or who knows what else!

Use the encyclopedia to begin your investigation into the history of the item, and outline your findings here!

_____
_____
_____
_____
_____
_____
_____
_____
_____
_____
_____
_____
_____
_____
_____
_____
_____
_____
_____
_____
_____
_____
_____

lever	rubber band	roller skates
button	magnifying glass	toothbrush
baseball	shoe lace	pencil
ball point pen	zipper	paper clip
	"sticky" notes	

Reference Resources
Inventions

Invention, by Lionel Bender. (Eyewitness Books). Grade 5 and up. Published by Dorling Kindersley, DK Family Learning, 7800 Southland Blvd, Suite 200, Orlando, FL 32809 (800) 352-6651, distributed by Houghton Mifflin, Industrial/Trade Division, 181 Ballardvale Rd., Wilmington, MA 01887 (800) 225-3362.

Story of Inventions, by Michael P. McHugh and Franck P. Bachman. Grades 4 - 8. Published by Christian Liberty Press. Available from Farm Country General Store, Rt. 1, Box 63, Metamora, IL 61548. (800) 551-FARM.

The Picture History of Great Inventors, by Gillian Clements. Published by Knopf Books for Young Readers, 400 Hahn Rd., Westminster, MD 21157. (800) 733-3000.

Machines & Inventions, (Understanding Science and Nature Series). Grades 7 and up. Published by Time Life Publishing, Dept. 100, Richmond, VA 23280. (800) 621-7026.

Invention & Discovery, by S. Reid. (Illustrated Dictionaries Series). Grades 6 and up. Published by EDC Publishing, 10302 E. 55th Place, Tulsa, OK 74146. Available from Great Christian Books, 229 S. Bridge St., P.O. Box 8000, Elkton, MD 21922. (800) 775-5422.

The Invention of Ordinary Things, by Don Wulffson. Grades 3 and up. Published by Lothrop, Lee and Shepard Books, William Morrow and Co., 39 Plymouth St., Fairfield, NJ 07007. (800) 237-0657.

Be An Inventor, by Barbara Taylor. Grades 3 - 8. (Weekly Reader Presents Series). Published by Harcourt Brace Jovanovich, 6277 Sea Harbor Dr., Orlando, FL 32886. (800) 782-4479.

Inventions have often been inspired when people wanted to find ways to do things faster, better, safer, or more conveniently. Sometimes, the inventor has had an idea for a completely different invention—a NEW thing or machine!

Use this space to develop, describe and illustrate YOUR idea for an invention!

Handier

Easier

Closer

Higher

Lower

Faster

Safer

Cooler

Warmer

Cheaper

Neater

Simpler

Cleaner

Wackier

Stronger

Better

Longer

Shorter

Softer

Taller

Easier

Slower

Boston's Museum of Science Inventor's Workshop, by Belinda Recio. (Discovery Kit Series). Grades 4 - 9. Published by Running Press Publishers, 125 South 22nd Street, Philadelphia, PA 19103.

Steven Caney's Invention Book, by Steven Caney. Grades 5 and up. Published by Workman Publishing Company, 708 Broadway, New York, NY 10003. (800) 722-7202.

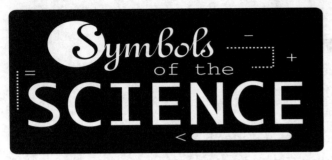

Symbols of the SCIENCE

Use this page to demonstrate your own circuit design. Don't forget to look up and use the right symbols for batteries, switches, etc. Think of designing a room alarm circuit, a light circuit, a new kind of toy buzzer, and who knows what else! Don't forget to remember the SAFETY rules we always use with electricity!

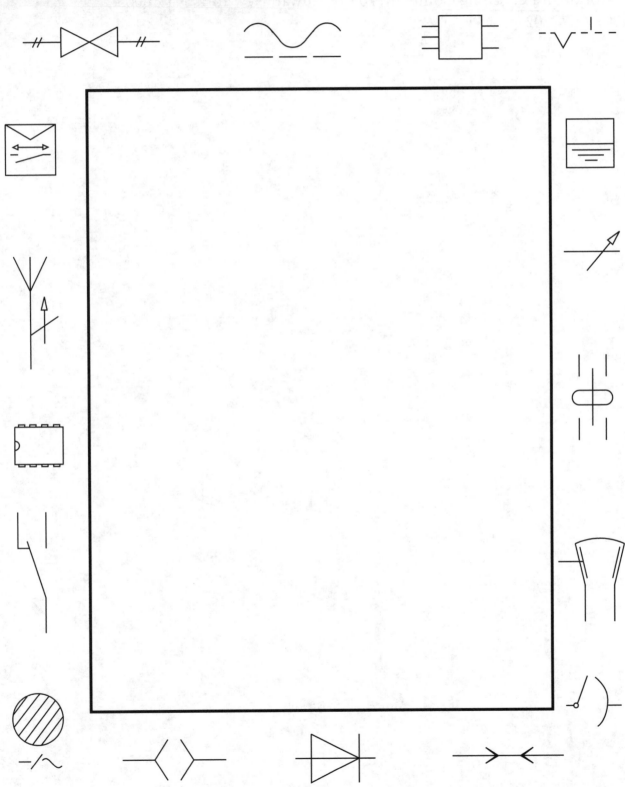

Reference Resources
Science

The Young Scientist Book of Electricity: Understanding the Secrets of Electric Power and How We Use It, by Philip Chapman. Grades 2 - 7. Published by EDC Publishing, 10302 E. 55th Place, Tulsa, OK 74146. (800) 475-4522.

Switch On, Switch Off, by Melvin Berger. (Let's Read and Find Out Series). Grades PreK - 3. Published by HarperCollins Children's Books, 1000 Keystone Industrial Park, Scranton, PA 18512. (800) 242-7737.

Understanding Electricity, by Gary Gibson. (Science for Fun Series). Grades 2 - 5. Published by Copper Beech Books, Millbrook Press, 2 Old New Milford Road, Brookfield, CT 06904-0335. (800) 462-4703.

Heat, Lights and Action! How Electricity Works, by Eve and Albert Stwertka. (At Home With Science Series). Grades 4 - 7. Published by Julian Messner, a division of Silver Burdett, Simon & Schuster, Inc., P.O. Box 2649, Columbus, OH 43216. (800) 848-9500.

Electronics, by Roger Bridgman. (Eyewitness Science Series). Grades 5 - 12. Published by Dorling Kindersley, DK Family Learning, 7800 Southland Blvd, Suite 200, Orlando, FL 32809 (800) 352-6651, distributed by Houghton Mifflin, Industrial/Trade Division, 181 Ballardvale Rd., Wilmington, MA 01887. (800) 225-3362.

Electricity and Magnetism, by Peter Adamczyk and Paul-Francis Law. (Usborne Understanding Science). Grades 5 and up. Published by EDC Publishing, 10302 E. 55th Place, Tulsa, OK 74146. (800) 475-4522.

What is Electronics, An electronics lab kit and detailed manual, for ages 10 and up. Available from Timberdoodle Company, E. 1510 Spencer Lake Rd., Shelton, WA 98584. (360) 426-0672.

Electricity and Magnetism, from the Milliken Science Series. Grades 5 - 8. Published by Milliken Publishing Company, P.O. Box 21579, St. Louis, MO 63132. (800) 325-4136. Available from Farm Country General Store, Rt. 1, Box 63, Metamora, IL 61548. (800) 551-FARM.

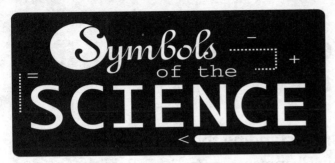

Use this page to demonstrate your own circuit design. Don't forget to look up and use the right symbols for batteries, switches, etc. Think of designing a room alarm circuit, a light circuit, a new kind of toy buzzer, and who knows what else! Don't forget to remember the SAFETY rules we always use with electricity!

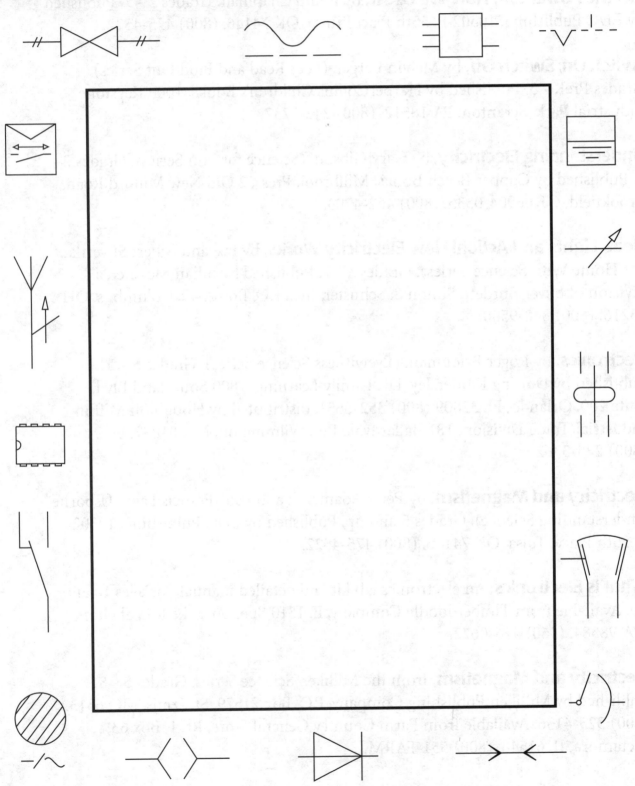

Wires & Watts: Understanding & Using Electricity, by Irwin Math. Grades 7 - 12. Published by Aladdin Books, Simon & Schuster, owners of Macmillan Publishing Company, 200 Old Tappan Rd., Old Tappan, NJ 07675. (800) 223-2336.

Where Does Electricity Come From?, by Susan Mayes. (Usborne Starting Point Science Series). Grades 1 - 4. Published by EDC Publishing, 10302 E. 55th Place, Tulsa, OK 74146. (800) 475-4522.

Superconductors: The Irresistible Future, by Albert Stwertka. Grades 7 - 12. Published by Franklin Watts, 5450 Cumberland Avenue, Chicago, IL 60656. (800) 672-6672.

More Wires & Watts: Understanding & Using Electricity, by Irwin Math. Grades 7 - 12. Published by Simon & Schuster Children's Books, 200 Old Tappan Rd., Old Tappan, NJ 07675. (800) 223-2348.

Exploring Electricity, by Ed Catherall. (Exploring Science Series). Grades 4 - 8. Published by Raintree Steck-Vaughn, P.O. Box 26015, Austin, TX 78755. (800) 531-5015.

Batteries, Bulbs and Wires: Science Facts and Experiments, by David Glover. (Young Discoverers Series). Grades 1 - 4. Published by Kingfisher Books LKC, Larousse, Kingfisher, Chambers, Inc., 95 Madison Avenue, New York, NY 10016. (800) 497-1657.

Discovering Electricity, by Rae Bains. Grades 2 - 4. Published by Troll Associates, 100 Corporate Dr, Mahwah, NJ 07430. (800) 929-8765.

Electricity & Magnets, by Barbara Taylor. (Science Starter Series). Grades 5 - 8. Published by Franklin Watts, 5450 Cumberland Avenue, Chicago, IL 60656. (800) 672-6672.

All About Radio and Television, by Jack Gould. Grades 4 - 8. (All About Series). Published by Random House, 400 Hahn Rd., Westminster, MD 21151. (800) 733-3000. An older book, © 1958, but presents a clear and simple explanation of the basics of radio and television in terms of electricity and operation.

Best Experiment BOOKS

Electricity is a fantastic topic to work on and learn about—using books as well as experiments! I've listed several experiment books to get you started, and I'm sure there are plenty more at your library.

Which ones were your favorites?

Why did you like them?

What were your favorite experiments?

wonder...view...attempt...decide...trial...check...test...measure...find...build...design...construct...safety...observe...ask...think...cause...effect...explore...succeed...complete...circuit...balance...increase...

Reference Resources
Experiments

Here are some books that contain all kinds of experiments with electricity, ranging from simple ones for the younger grades to more complex experiments for the upper grades. To assure everyone's safety, ALWAYS make sure that there is adult supervision of these experiments!

The Thomas Edison Book of Easy and Incredible Experiments: Activities, Projects and Science Fun for All Ages, by The Thomas Alva Edison Foundation. Grades 4 - 11. Published by John Wiley & Sons, Inc., 1 Wiley Drive, Somerset, NJ 08875. (800) 225-5945.

Fun with Electronics, by J.G. McPherson. (Pocket Scientist Series). Grades 3 - 6. Published EDC Publishing, 10302 E. 55th Place, Tulsa, OK 74146. (800) 475-4522.

Science With Magnets, **Science With Batteries** and **Energy & Power**, all from the Usborne Science Actvities Series. Grades 1 - 4. Published by EDC Publishing, 10302 E. 55th Place, Tulsa, OK 74146.

Young Scientist Book of Electricity. (Usborne Young Scientist Series). Grades PreK - 4. Published by EDC Publishing, 10302 E. 55th Place, Tulsa, OK 74146.

Science Book of Electricity, **Science Book of Energy** and **Science Book of Magnets**, all by Neil Ardley. Grades 3 - 6. (Science Book Series). Published by Harcourt Brace Jovanovich, 6277 Sea Harbor Drive, Orlando, FL 32886. (800) 782-4479.

TOPS Science Module: Electricity and **TOPS Science Module: Magnetism**, by Ron Marson. (TOPS Learning Systems Series). Grades 3 - 10. Published by TOPS Learning Systems. Available from Timberdoodle Company, E. 1510 Spencer Lake Rd., Shelton, WA 98584. (360) 426-0672.

Experimenting With Electricity and Magnetism, by Ovid K. Wong. Grades 7 - 12. Published by Franklin Watts, 5450 Cumberland Avenue, Chicago, IL 60656. (800) 672-6672.

Best Experiment BOOKS

Electricity is a fantastic topic to work on and learn about—using books as well as experiments! I've listed several experiment books to get you started, and I'm sure there are plenty more at your library.

Which ones were your favorites?

Why did you like them?

What were your favorite experiments?

wonder...view...attempt...decide...trial...check...test...measure...find...build...design...construct...safety...observe...ask...think...cause...effect...explore...succeed...complete...circuit...balance...increase...

Fun With Electronics: Build 25 Amazing Electronic Projects!, by Conn McQuinn. A complete kit that includes electronic components and wires, as well as a manual to guide the student through the projects. Grades 7 - 12. Published by Andrews and McMeel, Universal Press Syndicate Company, 4900 Main Street, Kansas City, MO 64112.

Fun With Electronics, Jr.: Hours of Experimenting Fun with a Buzzer, Motor, Lights, and More!, by Luann Colombo and Conn McQuinn. A complete kit that includes electrical components and wires, as well as a manual for the students and parent/teacher to use together. Grades 2 - 6. Published by Andrews and McMeel, Universal Press Syndicate Company, 4900 Main Street, Kansas City, MO 64112.

Power Up: Experiments, Puzzles & Games Exploring Electricity, by Sandra Markle. Grades 3 - 7. Published by Simon & Schuster Children's Books, 200 Old Tappan Rd., Old Tappan, NJ 07675. (800) 275-5755.

Janice VanCleave's Electricity: Mind-Boggling Experiments You Can Turn Into Science Fair Projects, by Janice VanCleave. (Spectacular Science Projects Series). Grades 3 - 6. Published by John Wiley & Sons, 1 Wiley Drive, Somerset, NJ 08875. (800) 225-5945.

Experiment With Magnets & Electricity, by Margaret Whalley. Grades 2 - 5. Published by Lerner Publications, 241 First Ave. N., Minneapolis, MN 55401. (800) 328-4929.

Electromagnetics in Action, by Parramon Staff. (Super-Charged Science Projects Series). Grades 5 and up. Published by Barron's Educational Series, Inc., 250 Wireless Blvd., Hauppauge, NY 11788. (800) 645-3476.

Famous Experiments & How to Repeat Them, by Brent Filson. Grades 5 and up. Published by Silver Burdett Press, subsidiary of Simon & Schuster, Inc., P.O. Box 2649, Columbus, OH 43216. (800) 848-9500.

Electronics Projects for Young Scientists, by George Leon. (Projects for Young Scientists Series). Grades 9 - 12. Published by Franklin Watts, 5450 Cumberland Avenue, Chicago, IL 60656. (800) 672-6672.

Favorite BIOGRAPHIES

A biography is a book that is written about someone's life. It provides a look "inside" a person's background, home life, education, and other areas that help us better understand the person that we are studying. There are many listed in this book to give you some ideas.

List the biographies that you enjoyed here, along with the author and a description of the book!

1. _____
_____
_____
_____

2. _____
_____
_____
_____

3. _____
_____
_____
_____

4. _____
_____
_____
_____

Who . . . what . . . where . . . when . . . why . . . Who . . . what . . . where . . . when . . . why . . . Who . . . what

where . . . when . . . why . . . Who . . . what . . . where . . . when . . . why . . . Who . . . what

Reference Resources
People

Michael Faraday: Father of Electronics, by Charles Ludwig. Grades 6 -10. Published by Herald Press, 616 Walnut Ave., Scottsdale, PA 15683. (800) 245-7894.

Thomas Edison and Electricity, by Steve Parker. (Science Discoveries Series). Grades 3 - 8. Published by HarperTrophy, HarperCollins Children's Books, 1000 Keystone Industrial Park, Scranton, PA 18512. (800) 242-7737.

Ben Franklin of Old Philadelphia, by Margaret Cousins. (Landmark Books Series). Grades 4 - 8. Published by Random House Books for Young Readers, 400 Hahn Rd., Westminster, MD 21157. (800) 733-3000.

The Story of Thomas Alva Edison, by Margaret Cousins. (Landmark Books Series). Grades 4 - 8. Published by Random House Books for Young Readers, 400 Hahn Rd., Westminster, MD 21157. (800) 733-3000.

Benjamin Franklin: Young Printer, by Augusta Stevenson. (Childhood of Famous Americans Series). Grades 2 - 6. Published by Simon & Schuster Children's Publishing Division, 200 Old Tappan Rd., Old Tappan, NJ 07675. (800) 223-2348.

Thomas Edison: Young Inventor, by Sue Guthridge. (Childhood of Famous Americans Series). Grades 2 - 6. Published by Simon & Schuster Children's Publishing Division, 200 Old Tappan Rd., Old Tappan, NJ 07675. (800) 223-2348.

Thomas Edison, by Kelly Anderson. Grades 5 - 8. Published by Lucent Books, P.O. Box 289011, San Diego, CA 92198. (800) 231-5163.

Young Tom Edison: Great Inventor, by Claire Nemes. (First-Start Biography Series). Grades K - 2. Published by Troll Associates, 100 Corporate Dr., Mahwah, NJ 07430. (800) 929-8765.

Thomas Edison: Great American Inventor, by Shelly Bedick. Grades PreK - 3. Published by Scholastic, Inc., P.O. Box 7502, Jefferson City, MO 65102. (800) 325-6149.

While there are many inventors that played key roles in the development of our use of electricity, there are several very important ones to investigate. Here are a few names—choose one and see what kind of interesting information you can find about this person! For example, did you know that Samuel Morse was at one time a portrait artist?

When and where was the inventor born? What kind of education did he receive? How did he get interested in inventing?

_____

_____

_____

_____

_____

_____

_____

_____

_____

_____

_____

_____

_____

_____

_____

_____

_____

_____

_____

_____

_____

_____

_____

Thomas Edison	Samuel Morse	Hans Christian Oersted
Benjamin Franklin	Alessandro Volta	Joseph Henry
James Watt	André-Marie Ampere	Alexander Bell
Luigi Galvani	Michael Faraday	Nikola Tesla
	George Ohm	

Samuel F. B. Morse: Artist With a Message, by John Tiner. (Sower Series).
Grades 3 - 8. Published by Mott Media, 1000 E. Huron, Milford, MI 48381.
(800) 421-6645.

Samuel F. B. Morse, by Jean Latham. (Discovery Biography Series). Grades 2 - 6.
Published by Chelsea House Publishers, 1974 Sproul Rd., Suite 400, P.O. Box 914,
Broomall, PA 19008. (800) 848-2665.

Guglielmo Marconi & Radio, by Steve Parker. (Science Discoveries Series).
Grades 3 and up. Published by Chelsea House Publishers, 1974 Sproul Rd.,
Suite 400, P.O. Box 914, Broomall, PA 19008. (800) 848-2665.

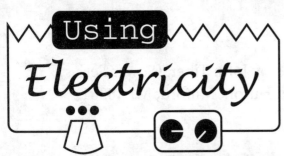

Using Electricity

Electricity is something we have to pay for—just like food, gasoline, and other things we use from day to day. The utility company charges for our monthly use based on how much electricity we have used, with a reading of each family's electric meter. With an adult, locate your meter (usually found outside of your home or apartment), and learn to take a reading from it.

Read your home's electric meter at the same time each day for one month. Record your findings here. Do you see a change in the amount of electricity your family uses on particular days? What are the "high use" days in your home?

Day	Reading	24-hr Total	Day	Reading	24-hr Total
___	_____	_____	___	_____	_____
___	_____	_____	___	_____	_____
___	_____	_____	___	_____	_____
___	_____	_____	___	_____	_____
___	_____	_____	___	_____	_____
___	_____	_____	___	_____	_____
___	_____	_____	___	_____	_____

Week 1 Total _____   Week 3 Total _____

Day	Reading	24-hr Total	Day	Reading	24-hr Total
___	_____	_____	___	_____	_____
___	_____	_____	___	_____	_____
___	_____	_____	___	_____	_____
___	_____	_____	___	_____	_____
___	_____	_____	___	_____	_____
___	_____	_____	___	_____	_____
___	_____	_____	___	_____	_____

Week 2 Total _____   Week 4 Total _____

NOTE: You will not have a 24-hour total for the first day of your log. To calculate the 24-hour total for the following day, subtract the previous reading from the most recent reading.

Reference Resources
Miscellaneous

Everyday Things & How They Work, by Peter Turvey. Grades 5 - 8. Published by Franklin Watts, 5450 Cumberland Ave., Chicago, IL 60656. (800) 672-6672.

The Way Things Work, by David Macaulay. Grades K - 12. Published by Houghton Mifflin, 181 Ballardvale Rd., Wilmington, MA 01887. (800) 225-3362.

Radio: From Marconi to the Space Age, by Alden Carter. Grades 4 - 9. Published by Franklin Watts, 5450 Cumberland Ave., Chicago, IL 60656. (800) 672-6672.

Career's Inside the World of Technology, by Jean Spencer. (Careers and Opportunities Series). Grades 7 and up. Published by Rosen Publishing Group, 29 E. 21st St., New York, NY 10010. (800) 237-9932.

Careers as an Electrician, by Elizabeth Lytle. Published by Rosen Publishing Group, 29 E. 21st St., New York, NY 10010. (800) 237-9932.

Careers in Electric Power, by Guy Carsone. Published by Edison Electric Institute, Washington, D.C. (202) 508-5000.

Exploring Careers as an Electrician, by Elizabeth Lytle. (Exploring Careers Series). Published by Rosen Publishing Group, 29 E. 21st St., New York, NY 10010. (800) 237-9932

SAVING ELECTRICITY

Now that you have learned to read the electric meter for your family's use of electricity, you can see just how much you depend on electricity! To save both electricity and money, there are all kinds of things we can do around the house that make a difference in how much we use.

What kinds of things can you and your family do to save electricity?
List them here, then try them out and watch the meter for a month or so.
What difference did the changes make?

Electricity • Electricity • Electricity • Electricity • Electricity • Electricity • Electricity • Electricity • Electricity • Electricity

Electricity • Electricity • Electricity • Electricity • Electricity • Electricity • Electricity

Working Outline

I. Introduction

A. Origin and meaning of the word "electric"

B. Electricity is one of the many forms of energy created by God that surround us

C. Electricity in our daily lives

    1. Our homes—electrical appliances such as lamps, air conditioning, stove, computer, etc.

    2. In the air—static electricity and lightning

    3. Our town—street lights, store signs, traffic lights, factories, stadium lights, etc.

D. Importance of electricity to people today

    1. Increases the quality of our lives, by powering improvements like central air conditioning and heating, better lighting, better food preparation and preservation equipment, etc

2. Improves our health standards, by providing better diagnostic equipment (X-ray, MRI, ultrasound) as well as better treatment options (lasers, endoscopic treatment, etc.)

3. Opens up the world around us, via television, multimedia sources, the Internet, etc., making people more aware of the needs, thoughts and actions of others

II. Science of electricity

A. Types of energy (examples)

1. Electricity

2. Heat

3. Light

4. Sound

B. Two types of electricity

1. Static electricity—electricity at rest

2.   Current electricity—electricity that travels through a complete circuit

   a.   Alternating Current (AC)—where the electron flow changes direction constantly

   b.   Direct Current (DC)—where the electron flow is only in one constant direction

C.   Electricity and atoms

   1.   The three main parts of an atom

      a.   Proton—positively charged particle

      b.   Electron—negatively charged particle

      c.   Neutron—neutral particle (no charge)

   2.   Electricity is the flow of electrons from atom to atom

      a.   Opposite charges attract

      b.   Like charges repel

D. Electrical flow through different materials

    1. Conductors

        a. Materials that allow a fast flow of electrons

        b. Examples include gold, silver, copper

    2. Insulators (nonconductors)

        a. Materials that resist or impede the flow of electrons

        b. Examples include wood, rubber, plastic

E. Circuits

    1. Defined as being a continuous or "closed" path for the flow of electricity

    2. Types of circuits

        a. Parallel—more than one path for the current to follow

        b. Series—only one path for the current to follow

3. Common components of a circuit (conductors)

    a. Wiring

    b. Power supply (example: battery)

    c. Switch

    d. Example of miscellaneous components—vary by the application of the circuit

        (1) Light bulb

        (2) Buzzer

        (3) Motor

        (4) Resistor

        (5) Capacitor

F. Units of measurement associated with electricity

    1. Ampere (amp)—measures current

2. Volt—measures potential energy

3. Ohm—measures resistance to electrical flow

4. Watt—measures power

III. History, scientists, and inventors

A. Ancient history of experimenting with electricity

1. Thales of Miletus (625 - 547 B.C.)

a. Greek mathematician and philosopher

b. Discovered that amber stone (fossilized tree resin) attracted lightweight objects after being rubbed

2. Pliny the Roman (First century A.D.)

a. Statesman and writer

b. Studying Thales' findings, Pliny performed some experiments of his own

        c.   Wrote about Thales' work and his own experiments

B.   Renewed interest in electricity, 17th-18th centuries

    1.   William Gilbert

        a.   Physician to Queen Elizabeth I of England (1544-1603)

        b.   In 1600, he wrote about his experiments with magnets and the attraction of various materials after being rubbed

        c.   Named the materials that attract other materials "electrics" after the Greek word for amber, elektron

        d.   Explained how the earth itself acts as magnet, offering an explanation of how the compass works

    2.   Otto Von Guericke (1602-1686)

        a.   German mayor of the town of Magdeburg

        b.   Invented the first friction machine, used to generate electricity

c. Constructed his friction machine from a ball made of sulfur, rubbing the ball with a cloth produced sparks of electricity

3. Stephen Gray (1666-1736)

   a. An English scientist, experimented in the 1720s with the movement of electricity, and found that electricity could move from one object to another

   b. Discovered that electricity moves easily through some materials, which he called conductors

   c. Materials that electricity does not flow through (silk, glass, etc.) were called non-conductors, which lead to their use as insulators to control the flow of electricity

4. Charles Francis DuFay (1698-1739)

   a. French scientist, experimented in 1733 with electricity and found that there were two different kinds of electricity

      (1) Static electricity, which is generated from rubbing two objects together

      (2) Current electricity, where electricity is actually moves through a material (electron flow)

5. In 1746, the Leyden jar was developed by scientists at the University of Leyden in the Netherlands, and used to store electrical charge

6. Benjamin Franklin (1706-1790)

    a. American scientist, inventor and statesman

    b. Experimented with the behavior of electricity, discovering and naming objects as positively or negatively charged

    c. Proved that lightning was actually an electric spark with his famous kite experiment in 1752

    d. Invented the lightning rod for buildings, to prevent their destruction from lightning

7. Luigi Galvani (1737-1798)

    a. Italian physician and professor

    b. Discovered in 1771 that frog legs used for an experiment would twitch when they came in contact with two metals, calling this electricity "animal electricity"

8. Alessandro Volta (1745-1827)

    a. Italian history professor, disagreed with Galvani's concept of "animal electricity"

    b. Worked with metals to develop the "voltaic pile", which was composed of various metal discs as well as other materials

    c. The voltaic pile produced a continuous electrical flow—the first battery

    d. The unit of measurement of electric potential, volt, is named after Volta

C. Discoveries in the nineteenth century and beyond

    1. Hans Christian Oersted (1777-1851)

        a. Professor of physics, University of Denmark

        b. In 1820, he discovered that magnetism was produced by an electric current

        c. The operation of the electric motor is based on his theory of electromagnetism

2. André-Marie Ampère (1775-1836)

    a. French scientist, experimented and discovered ways to mathematically explain the magnetic fields generated by electric current

    b. The unit of measurement of electrical current is called the amp, named in his honor

3. Georg Simon Ohm (1787-1854)

    a. German scientist, discovered that all materials have some form of resistance to electrical current—called this the "resistance" of the material, later to be measured in units of ohms, named in his honor

    b. Developed the mathematical relationship between resistance and current and electrical potential, called Ohm's law

4. Michael Faraday (1791-1867)

    a. English scientist, called the "Father of Electricity" because of his discovery of the basics of electromagnetic induction, the principle of the electric motor

    b. Invented the modern electric generator (dynamo)

5. Samuel F. Morse (1791-1872)

   a. American artist, designed an electromagnetic telegraph and a code for sending messages on his telegraph (Morse code)

   b. In 1844, the first permanent telegraph transmitted Morse's famous first message, from Washington to Baltimore

6. Joseph Henry (1797 - 1878)

   a. American scientist, noted for his work with electromagnetics

   b. Discovered the principle of self-induction, leading the way for the development of efficient electric motors

7. Alexander Graham Bell (1847-1922)

   a. Born in Scotland, and later became a United States citizen, invented the first practical telephone

   b. Worked on the study of sound and educating the deaf

8. Thomas Edison (1847-1931)

    a. Brilliant American inventor of both electrical and mechanical apparatus

    b. Held over 1,000 patents in his lifetime

    c. Opened the first power station

    d. Some of his greatest inventions include the electric light bulb, the phonograph, and the kinetoscope

9. Nikola Tesla (1856 - 1943)

    a. Engineer and inventor, native of Croatia, later emigrated to America

    b. Remembered for his discovery of alternating current as well as his work with motors

IV. Applications of electricity

A. Residential

    1. Comfort

        a. Air conditioning and heating systems

        b. Lighting for night time comfort and tasks

        c. Ceiling fan

        d. Security system

    2. Household responsibilities

        a. Clothing care and construction

            (1) Sewing machine

            (2) Washing machine

            (3) Clothes dryer

            (4) Iron

b.   Food preparation

    (1)   Refrigerator/freezer

    (2)   Oven

    (3)   Range

    (4)   Microwave

    (5)   Dishwasher

    (6)   Small appliances like can openers, bread machines, toasters, etc.

c.   Home care and cleaning

    (1)   Vacuum cleaner

    (2)   Hot water heater

d.   Exterior home care and repair

    (1)   Power tools

        (a)   Drill

        (b)   Saw

        (c)   Compressor

(2) Lawn care

    (a) Electric lawn mower

    (b) Hedge trimmer

    (c) Edger

3. Home business and recreation

    a. Computer equipment

    b. Fax machine

    c. Telephone equipment

    d. Television and VCR

B. Agricultural applications

    1. Food preparation equipment

    2. Animal care equipment

    3. Crop care equipment

C. Commercial applications (stores, banks, restaurants, etc.)

    1. Air conditioning and heating systems

    2. Food preparation equipment

    3. Business equipment

        a. Computer equipment

        b. Cash register

        c. Copy machine

        d. Fax machine

        e. Adding machine

    4. Security system

D.   Industrial applications (manufacturers)

    1.   Production equipment

    2.   Inventory management equipment

    3.   Business equipment

    4.   Comfort equipment

    5.   Safety equipment

V.   Impact of innovations and inventions that utilize electricity over the last 150 years

A.   Economic impact

    1.   Major change in industrial efficiency and productivity resulted from the use of electricity

    2.   As industry changed and became more productive, economies grew and strengthened

3. With the growing economies came more investment in industry, resulting in new interest in research and development to find new products

B. Worldwide changes

1. Electricity is now produces around the world

2. Products can now be manufactured in areas that previously would not have been considered

3. Widespread use of these innovations has transformed the world in how we live, work, and play

VI. Electricity today—the production and transmission by utility companies

A. Electricity is now generated by major utility companies in several ways

1. Fossil fuel plants (coal-fired)

2. Nuclear plants

3. Hydro-electric plants (using dams)

4. Alternative sources

    a. Solar equipment

    b. Wind turbines

B. Transmitting the electricity from the power plant to your home

    1. Electricity is transmitted through what is known as "the grid"—a network of cables connected throughout the country for the sole purpose of delivering electricity

    2. From the power plant to your home, the electricity will go through several sets of transformers to bring the power to residential requirements (110 Volts)

    3. At your home, the power will come through a fuse box or breaker box, delivering power to the various areas or zones of your home

VII. The future holds bright promise and many possibilities for new and improved inventions as well as applications of electricity in many areas of our lives

A. Medicine

    1. Better scanning methods

2.   Enhanced medication delivery systems (shots that won't hurt!)

3.   Improved methods of surgery

B.   Electronics

    1.   Increased efficiency

    2.   Larger memory on smaller chips

    3.   Faster operating systems that use less electricity

C.   Communications

    1.   Smaller and more affordable telephones using cellular or other satellite technology

    2.   Improved satellites for communications

    3.   Effective ways to utilize the Information Highway

D. Education

    1. Improved computers and software systems to increase educational options

    2. Virtual technology applications in education

    3. Online books and resources will become more readily available using smaller and more efficient satellite technology and cable systems

E. Transportation

    1. Improved electric cars

    2. New ways to utilize electricity in mass transit

    3. Improved systems and methods of propulsion and energy for the space program

F. Government

    1. Improved voting equipment and counting systems

2.   With the vast media information becoming more quickly and readily available, accountability of our elected officials should improve

3.   Better use of computers and new scanning and reporting systems will keep the public much more aware of government operations

Finish Line FUN!

I always try to complete a unit study with some kind of finale—a grand finish which is fun and rewarding and also showcases the children's efforts. Picnics, museum trips, pizza parties, and family weekends are some good ways to accomplish this.

Plan the framework of your "finish line fun" here, and then build toward it during your study!

Electricity • Electricity

About The Author

Amanda Bennett, author and speaker, wife and mother of three, holds a degree in mechanical engineering. She has written this ever-growing series of unit studies for her own children, to capture their enthusiasm and nurture their gifts and talents. The concept of a thematic approach to learning is a simple one. Amanda will share this simplification through her books, allowing others to use these unit study guides to discover the amazing world that God has created for us all.

Science can be a very intimidating subject to teach, and Amanda has written this series to include science with other important areas of curriculum that apply naturally to each topic. The guides allow more time to be spent enjoying the unit study, instead of spending weeks of research time to prepare for each unit. She has shared the results of her research in the guides, including plenty of resources for areas of the study, spelling and vocabulary lists, fiction and nonfiction titles, possible careers within the topic, writing ideas, activity suggestions, addresses of manufacturers, teams, and other helpful resources.

The science-based series of guides currently includes the Unit Study Adventures titles:

Baseball	Oceans
Computers	Olympics
Elections	Pioneers
Electricity	Space
Flight	Trains
Gardens	Dogs
Home	

The holiday-based series of guides currently includes the Unit Study Adventures titles:

Christmas
Thanksgiving

This planned 40-book series will soon include additional titles, which will be released periodically. We appreciate your interest. "Enjoy the Adventure."